GEMINI 2.0
Everything You Need to Know

How Google's Latest AI Revolution is Redefining Smarts, Speed, and Solutions

J. Andy Peters

Copyright ©**J. Andy Peters, 2024.**

All rights reserved. No part of this publication may be reproduced, distributed, or transmitted in any form or by any means, including photocopying, recording, or other electronic or mechanical methods, without the prior written permission of the publisher, except in the case of brief quotations embodied in critical reviews and certain other noncommercial uses permitted by copyright law.

Table of Contents

Introduction 3
Chapter 1: The Genesis of Gemini 2.0 6
Gemini 1.0 and 1.5: Expanding Multimodal AI Possibilities 9
The Leap from Gemini 1.5 to 2.0: A Quantum Jump Forward 12
Chapter 2: A Leap in Multimodal AI 16
Chapter 3: Smarter, Faster, and More Accurate 23
Chapter 4: Going Beyond Traditional AI 33
Project Astra: A Universal Assistant with Memory and Context 38
Project Mariner: Navigating the Browser for Real-Time Task Completion 41
Chapter 5: AI in the Real World – Transforming Industries 49
Chapter 6: Ethical Considerations and Safety Measures 59
Chapter 7: The Competitive Landscape 71
Chapter 8: What's Next for Gemini 2.0 and AI? 85
Conclusion 99

Introduction

In the ever-evolving world of artificial intelligence, every new advancement feels like a giant leap forward. But few releases have captured the collective imagination like Google's Gemini 2.0. This isn't just another update; it's a game-changer, pushing the boundaries of what AI can achieve. In a world where we are constantly surrounded by new technologies, Gemini 2.0 stands out not only because of its incredible capabilities but also because of its potential to transform how we interact with the digital world. From solving complex problems to generating multimodal content, the Gemini 2.0 update promises to redefine what's possible.

For those of us who've been following the rapid progression of AI over the years, it's clear that we are standing on the edge of something monumental. And that's precisely where Gemini 2.0 comes in. Google's latest AI release marks a turning point, positioning us closer to a future where machines don't just assist us—they think, reason, and act with us. In this book, we will explore Gemini 2.0 in depth, breaking down its features and potential in a way that makes sense both for tech enthusiasts and everyday users alike. You don't need to be a developer to understand how this technology will affect your daily life—whether it's through the way you search the web, interact with devices, or even how you conduct business.

As we delve into the heart of this technology, the question of *What exactly is Gemini 2.0?* will be answered. At its core, Gemini 2.0 is an advanced AI model capable of processing both multimodal inputs and outputs. What does that mean? In simpler terms, it can understand and generate not just text, but also images, audio, and even video. It's faster, smarter, and more capable than anything that's come before. This update is more than just a tool—it's a powerful system designed to reshape entire industries, from software development and research to entertainment and education.

But what makes Gemini 2.0 particularly significant isn't just its technology—it's the way it integrates seamlessly into everyday life. Its impact won't be limited to labs or developer environments; this is a tool that's going to be as useful for everyday tasks as it is for complex coding. In the following pages, we'll take a closer look at how Google's Gemini 2.0 is reshaping the AI landscape and, more importantly, how it will reshape the way we live and work in a world increasingly driven by artificial intelligence.

Chapter 1: The Genesis of Gemini 2.0

The journey of Google's Gemini series began as an ambitious project aimed at transforming the AI landscape. In the world of artificial intelligence, Google had already made waves with their impressive advancements in machine learning and natural language processing. But the Gemini series was something different—it was designed to take AI to a new frontier, where it could think, reason, and create with human-like capabilities. To understand the significance of Gemini 2.0, it's essential to first look at where it all began: Gemini 1.0.

Gemini 1.0 was a groundbreaking release that marked a pivotal shift in Google's approach to AI. Before Gemini, AI models were powerful, but their capabilities were often confined to specific types of tasks. You had AI models that could process text, others that could handle images, and still others that were specialized in understanding and generating audio. However, these systems lacked the ability to integrate these different modalities in a meaningful way. Gemini 1.0 changed that by being the first truly multimodal AI model. It wasn't just about text generation anymore; it could process and understand not only written language but also images, videos, and even code. This was a huge leap forward, opening up a range of new possibilities for developers and users alike.

What made Gemini 1.0 so revolutionary wasn't just its ability to handle different types of input, but its capacity to combine them in ways that hadn't been done before. For example, it could analyze an image and generate a detailed written description of it, or take a video and produce a summarizing text. This integration of multiple inputs and outputs allowed Gemini to function in a far more sophisticated and holistic manner than any AI system that came before it. It marked the beginning of what Google referred to as the "agentic era"—an era in which AI wasn't just a tool that followed commands, but something capable of actively helping users achieve their goals.

While Gemini 1.0 was impressive, it was clear that it was just the beginning. Google had laid the foundation, and the potential for growth was limitless. Developers quickly recognized the power and versatility of Gemini 1.0, and the demand for more advanced capabilities only grew. This demand set the stage for the next evolution in the Gemini series: Gemini 2.0. The leap from 1.0 to 2.0 wasn't just an incremental update—it was a bold step forward that would push the boundaries of what AI could do, turning the promise of a truly intelligent, multimodal assistant into a reality.

Gemini 2.0 builds on the success of 1.0, refining and expanding its capabilities in ways that were previously unimaginable. Where Gemini 1.0 was a powerful introduction to multimodal AI, 2.0 is a more mature and fully realized version, ready to tackle the complex challenges of tomorrow. This evolution wasn't just about improving speed or accuracy—it was about fundamentally rethinking what an AI model could do. With the foundation laid by Gemini 1.0, Gemini 2.0 took those initial steps and expanded them into something much bigger. And as we will see, this next chapter in AI evolution is just the beginning.

As the Gemini series evolved, it marked a steady and powerful march toward a future where AI could understand and interact with the world in increasingly complex and meaningful ways. After the launch of **Gemini 1.0**, which set the stage for multimodal AI, Google quickly realized that the potential for such technology was enormous, and there was much room for growth. The company's focus turned toward refining this promising model, culminating in the release of **Gemini 1.5**, which built upon the foundation laid by its predecessor.

Gemini 1.0 and 1.5: Expanding Multimodal AI Possibilities

Gemini 1.0's groundbreaking success lay in its ability to process and understand various forms of data—text, images, videos, and even code. But despite its impressive capabilities, it was clear that there was still a lot more to be done. The first version allowed AI to understand and respond to these different types of inputs, but it was largely focused on text-based outputs. For instance, it could generate descriptive text based on an image or video, but it couldn't produce outputs that combined text with other modalities, such as creating multimedia responses that included both images and text.

With **Gemini 1.5**, Google made a significant refinement. While still leveraging the multimodal framework of 1.0, this version began shifting towards not just processing multiple forms of input but also producing more advanced multimodal outputs. It represented a pivotal evolution from simply being an input-based system to an output-driven one. Now, Gemini could not only understand text, images, and videos but also generate a wider array of outputs that integrated multiple modalities. For example, it could take an image and generate an appropriate script or combine different elements like text, audio, and visuals in a cohesive way, significantly improving the AI's creative and functional potential.

The development of **early developer tools** was another key aspect of Gemini 1.5's progression. Google introduced new APIs and resources that enabled developers to experiment with the model more easily. Tools like **Google AI Studio** and **Vertex AI** allowed developers to tap into the power of Gemini 1.5 and integrate these advanced capabilities into their own applications. This early access to such tools sparked a wave of innovation, with developers testing Gemini's ability to power everything from virtual assistants to creative applications, laying the groundwork for future iterations of the model.

The Leap from Gemini 1.5 to 2.0: A Quantum Jump Forward

However, it was with the release of **Gemini 2.0** that Google truly pushed the boundaries of what AI could achieve. While 1.0 and 1.5 made great strides in multimodal AI, Gemini 2.0 represents a leap forward in several key areas: **speed**, **reasoning**, and **multimodal capabilities**.

Speed was one of the most immediately noticeable improvements in Gemini 2.0. The model could now process tasks twice as fast as its predecessor, while maintaining—or even improving—its accuracy. This was a game-changer for developers, who now had an AI that could handle complex queries and tasks at an unprecedented pace. This improvement in processing speed not only made Gemini 2.0 more efficient but also more versatile, as it could tackle a broader range of real-time tasks without slowing down.

But speed alone wasn't enough. **Reasoning** and **problem-solving** were also significantly improved in Gemini 2.0. The AI model wasn't just about providing information or responding to simple queries—it was now capable of solving complex problems with better planning and advanced logic. This shift in how the AI reasoned through problems gave it a distinct edge, especially for more sophisticated use cases such as advanced research, coding, and business intelligence. The model could now process intricate instructions and generate deeper, more insightful responses, making it far more effective for both technical and everyday tasks.

Perhaps the most exciting advancement in Gemini 2.0, however, was in its **multimodal capabilities**. While earlier versions could process multiple types of input, it was in Gemini 2.0 that the AI truly shone in creating outputs that reflected this multimodal integration. It could now **generate images**, produce **multilingual**

speech, and even create **audio** outputs seamlessly. For example, if a user described a scenario, Gemini 2.0 could not only generate a written script but also provide accompanying visuals, audio cues, or even video clips to bring the description to life. This seamless creation of multimodal outputs represents a bold step into the future of AI interaction, where the technology can produce complex, dynamic content in real-time—content that's not just reactive, but also proactive and intuitive.

In addition to these core improvements, **Gemini 2.0** incorporated a more **robust API system** for developers, allowing them to harness the full power of the model for even more dynamic, interactive applications. From gaming to web navigation, Gemini 2.0's integration with tools like **Google Search** and **Google Maps** brought a level of smart, intuitive interaction that felt almost human. For example, the model could help users plan a trip by suggesting routes, providing localized information, and even answering follow-up questions—all without ever losing the context of the ongoing conversation.

The leap from Gemini 1.5 to 2.0 wasn't just a refinement of existing features; it was a reimagining of what AI could do. By improving speed, deepening reasoning capabilities, and enhancing multimodal outputs, Gemini 2.0 offered a more powerful and sophisticated toolset for developers, researchers, and everyday users. Its ability to process and create across multiple modalities, combined with faster response times and more insightful reasoning, makes it one of the most exciting advancements in the world of artificial intelligence. The next chapter of AI is here, and it's embodied in the power and potential of **Gemini 2.0**.

Chapter 2: A Leap in Multimodal AI

In the world of artificial intelligence, the concept of "multimodal" refers to the AI's ability to process and interpret multiple types of data at once, such as text, images, video, and audio. This is a significant leap from earlier AI models, which were primarily designed to handle one form of input at a time, typically text. Imagine, for a moment, an AI that can analyze a picture, understand the text embedded in the image, and even generate a corresponding video or audio description, all in one go. That's the power of multimodal AI. It's about breaking down barriers between different kinds of information and synthesizing them into a coherent understanding or output.

Gemini 2.0 takes this concept of multimodal AI to a whole new level. While earlier versions could only interpret various forms of input—such as processing images or reading text—the latest update goes further by enabling the AI to not only understand but also create these different outputs.

It's one thing to recognize an image, but Gemini 2.0 can now generate high-quality images from scratch based on a description. It can listen to text or voice commands in multiple languages and then produce accurate, multilingual speech outputs. And it doesn't stop there. Gemini 2.0 can combine different modalities to create dynamic content—combining text with visuals or generating video along with audio, offering a more immersive and interactive experience for users.

This expanded ability to generate multimodal outputs opens up countless possibilities, making Gemini 2.0 far more versatile and capable than its predecessors. Developers, for example, can use Gemini 2.0 to generate everything from complex images for creative projects to educational content that combines text explanations with illustrative diagrams or video clips. But the impact goes beyond just creativity. In fields like research, marketing, or even customer service, the ability to provide real-time, multimodal responses to queries elevates the user experience to a new level of convenience and depth.

This jump in capabilities doesn't just represent an incremental upgrade; it's a defining moment for AI as we move toward a future where machines can intuitively interact with us in ways that feel more natural and human-like. It's not about just reading or listening anymore—it's about seeing, hearing, and creating, all in a seamless dance of data that feels more like human interaction than ever before. And this is just the beginning. As Gemini 2.0 continues to evolve, we can only imagine what new forms of multimodal integration will emerge, reshaping industries and daily life in unexpected ways.

The ability of Gemini 2.0 to generate multimodal outputs opens up a world of practical applications that were once thought to be the stuff of science fiction. Think about the complexities of modern communication and the diverse forms in which information is consumed today—images, video, text, and audio all play an essential role. With Gemini 2.0, AI is now capable of creating a more complete and richer form of content.

For example, imagine a business that needs to produce marketing materials for a new product launch. Instead of hiring an entire team to create content in various formats, a single prompt to Gemini 2.0 could result in a fully formed video script, combined with a set of high-quality images to match, and an engaging voiceover in multiple languages. The AI can take the textual description of the product and transform it into everything needed for an entire marketing campaign—complete with visuals, sound, and a cohesive narrative. The script can be designed to capture the attention of diverse audiences, with the language automatically translated and tailored to fit local dialects. The ability to generate this type of content quickly, efficiently, and in multiple formats makes Gemini 2.0 a powerful tool for businesses looking to streamline their content creation processes.

Gemini 2.0's applications extend well beyond marketing. Customer service is another area where multimodal AI can make a profound impact. Imagine a customer interacting with a chatbot to resolve an issue. Traditionally, chatbots are limited to text-based responses. However, with Gemini 2.0, the AI can generate real-time, human-like responses through text, voice, and even video. This creates a more engaging and interactive experience for the customer. For example, if a user needs technical support, the AI could walk them through troubleshooting steps with both visual cues on a screen and spoken instructions in their native language, providing a truly personalized and intuitive customer service experience. This could significantly improve satisfaction levels while reducing the need for human intervention in many cases.

On the developer side, Gemini 2.0 has introduced a set of tools that empower creators to experiment and integrate these advanced functionalities into their applications. Google's Gemini API, available on platforms like Google AI Studio and Vertex AI, provides developers with the ability to harness the power of multimodal AI and customize it for specific use cases. For instance, through the Gemini API, developers can integrate AI-driven content generation tools into their websites, mobile apps, or even enterprise systems, unlocking a new level of interaction and automation. Whether it's creating dynamic user interfaces, building automated customer support agents, or generating diverse forms of content for digital platforms, the possibilities are endless.

For those working on large-scale applications, Gemini 2.0's integration into Google Cloud services like Vertex AI further enhances its accessibility and scalability. This means developers can tap into the computational power and storage resources provided by Google, enabling them to scale multimodal AI solutions to meet the needs of businesses and organizations, regardless of size.

In essence, Gemini 2.0's advanced capabilities offer not only immediate benefits for end-users in terms of enhanced interaction and content generation but also invaluable tools for developers, fostering creativity and efficiency in building next-generation applications. The seamless integration of text, image, video, and audio into a single AI framework promises to unlock a new era of possibilities across industries—driving innovation and reshaping how technology interacts with us on a day-to-day basis.

Chapter 3: Smarter, Faster, and More Accurate

The advancements made in Gemini 2.0 extend far beyond just adding new features; they include significant improvements in the AI's underlying performance, making it not only faster but also more accurate and reliable. In a landscape where speed and precision are key, Gemini 2.0 has taken monumental strides to ensure that it outperforms its predecessors.

One of the most notable enhancements in Gemini 2.0 is its processing speed. The model is capable of completing tasks at double the speed of the previous version, all while maintaining high levels of accuracy. This improvement is particularly critical for developers and businesses that rely on AI to handle complex, time-sensitive tasks. Whether it's handling real-time data processing, responding to queries, or generating content, Gemini 2.0 ensures that tasks are completed much faster, reducing latency and providing an overall smoother experience.

Beyond speed, Gemini 2.0 also excels in accuracy, particularly when it comes to understanding and solving complex problems. This is not just about processing text or images more efficiently; it's about providing the correct answers with a level of precision that is necessary for high-stakes applications. For example, in areas like data analysis, engineering, and technical troubleshooting, Gemini 2.0's ability to quickly identify and solve problems ensures that users get the right solutions the first time, every time. This type of accuracy boosts the reliability of the AI, which is crucial for adoption in industries that demand high levels of precision, such as healthcare, finance, and research.

One of the standout features of Gemini 2.0's performance is its exceptional achievement in code generation. Gemini 2.0 scored an impressive 90.2% on the Natural Language to Code benchmark, setting a new bar for AI in the realm of software development. This benchmark evaluates how well an AI can translate natural language instructions into functional, syntactically correct code. Prior to Gemini 2.0, this was a task that could challenge even advanced AI models, but the latest version has shown it can handle even the most complex coding queries with remarkable efficiency.

This leap forward is significant for developers, who often face the challenge of translating vague or high-level project descriptions into code. With Gemini 2.0, developers now have an AI that not only understands the intricacies of programming languages but also grasps the underlying logic of the task at hand. The AI can generate code that is not only accurate but also optimized for performance, saving developers both time and effort. This means that software projects, especially those with intricate coding requirements, can be completed faster and with fewer errors.

Furthermore, Gemini 2.0's improvements in code generation are not limited to basic code-writing tasks. The model's capabilities extend to more sophisticated tasks such as debugging, code optimization, and even suggesting alternative solutions for complex problems. In real-time development environments, Gemini 2.0 can assist developers by analyzing the code they write, identifying potential issues, and proposing fixes or improvements. This level of support is especially useful when working on large-scale or high-visibility projects where every line of code counts.

In addition to the technical advancements, Gemini 2.0's ability to provide rapid, accurate code generation can significantly speed up the software development lifecycle. This means faster prototyping, quicker debugging, and more efficient collaboration among development teams. For companies and organizations that rely on software to drive their business, Gemini 2.0's ability to streamline the coding process can lead to quicker turnaround times, faster product releases, and ultimately, a competitive edge in the market.

In essence, Gemini 2.0's performance boosts in speed, accuracy, and code generation represent not just incremental improvements but a fundamental shift in the way AI can be utilized across industries. Its ability to handle both general and technical tasks with greater efficiency and precision marks a new chapter in AI's evolution, making it a far more valuable tool for developers and businesses looking to harness the full potential of artificial intelligence.

Gemini 2.0's impressive speed and efficiency don't just look good on paper – they translate directly into real-world benefits for developers and everyday users. As tasks become more complex and the demands on artificial intelligence grow, the ability to process and deliver results quickly without compromising on accuracy is more important than ever. Gemini 2.0 has been designed with this in mind, offering a seamless experience for those who rely on its power for everything from quick answers to intricate problem-solving.

For developers, this means a drastic reduction in task completion time. In the past, processing large datasets or running complex queries could take considerable time, slowing down workflows and creating bottlenecks. With Gemini 2.0, tasks that once took minutes or hours to complete can now be executed in a fraction of the time. This speed not only helps developers get more done in less time, but it also enables them to iterate faster, making it easier to refine their work and test new ideas. Whether it's generating code, solving technical problems, or processing complex data sets, the AI's ability to keep up with the pace of development enhances productivity and accelerates project timelines.

For end users, Gemini 2.0's rapid processing is a game-changer. Tasks like searching for information, troubleshooting technical issues, or interacting with AI-driven tools become more efficient, with results delivered almost instantly. As the AI learns to understand and solve queries faster, users benefit from a more fluid, responsive experience. Whether you're using it to research a topic, plan a project, or get answers to detailed questions, the speed with which Gemini 2.0 responds ensures that you can get more done in less time, enhancing both individual productivity and overall user satisfaction.

The key to Gemini 2.0's speed and efficiency lies not just in its software, but also in the hardware that powers it. Google's Trillium TPUs (Tensor Processing Units) play a pivotal role in delivering the raw computational power necessary to handle the heavy workloads that come with AI tasks. These cutting-edge chips are designed specifically for the type of machine learning workloads that Gemini 2.0 demands, enabling it to process vast amounts of data at lightning speed while maintaining top-tier accuracy.

Trillium TPUs are built to optimize the performance of AI models by speeding up both training and inference processes. When training Gemini 2.0, these TPUs enable the AI to learn from massive datasets much more quickly than would be possible with traditional hardware. This results in a more refined and capable model, able to understand and generate text, images, audio, and other data formats in record time. In real-time applications, such as code generation or multimodal output, Trillium TPUs ensure that the AI can respond to requests without any noticeable delay, providing a smooth user experience.

One of the standout features of Gemini 2.0 is that these Trillium TPUs are not confined to Google's internal use. Google has made them available to customers, allowing businesses, developers, and other users to harness the same power that drives Gemini 2.0. This means that companies can leverage the full potential of the AI model, backed by the raw processing capability of Trillium TPUs, to supercharge their operations. Whether it's for research and development, customer support, or even large-scale data processing, users can tap into this advanced hardware to execute tasks faster and more efficiently.

For developers, this opens up a wealth of possibilities. By utilizing Trillium TPUs, they can scale their applications and services with the confidence that they have the processing power needed to handle the most demanding AI tasks. The hardware's ability to optimize performance means that developers can experiment with new features, push the boundaries of what's possible, and quickly deploy solutions without worrying about performance limitations. Additionally, the availability of TPUs through Google Cloud makes it easier for organizations of all sizes to access and integrate this powerful technology into their workflows.

In essence, the combination of Gemini 2.0's software capabilities and the support of Google's

Trillium TPUs creates an ecosystem that is primed for speed, efficiency, and scalability. As developers and users alike continue to explore the potential of this powerful AI model, the ability to complete tasks faster, with more accuracy, and with greater ease will become a defining factor in the way AI is integrated into both technical and everyday applications. The speed and processing power offered by Gemini 2.0, enhanced by Trillium TPUs, ensure that it remains at the forefront of AI innovation, delivering results that are not only accurate but also delivered at a pace that meets the demands of the modern world.

Chapter 4: Going Beyond Traditional AI

Agentic AI represents a significant leap forward in artificial intelligence, pushing the boundaries of what AI can do beyond simply responding to queries or providing information. While traditional AI models like chatbots or search engines are designed to answer questions and assist with specific tasks, Agentic AI takes things a step further by enabling machines to take independent action. With Gemini 2.0, this concept is brought to life in powerful ways that transform how AI interacts with users and systems.

At its core, Agentic AI refers to artificial intelligence that possesses not only the ability to process information but also the capacity to make decisions and take action based on that information. It involves planning, strategizing, and carrying out tasks autonomously, without the need for direct human intervention at each step. This capability allows AI to act as more than just a tool—it's a proactive agent that can handle complex, dynamic tasks with minimal input from users.

Gemini 2.0 takes full advantage of the principles of Agentic AI, allowing it to go beyond just answering questions or solving problems reactively. The system is designed to understand goals, devise a plan, and take the necessary steps to achieve those goals, whether it's generating content, solving a complex technical problem, or automating a multi-step process.

For instance, imagine an AI tasked with optimizing a website's SEO performance. In a traditional setup, a user might need to ask the AI for recommendations, then manually implement those suggestions. With Gemini 2.0's Agentic AI capabilities, however, the system could go beyond merely offering suggestions—it could autonomously assess the current state of the website, formulate an optimization strategy, make the required changes, and even monitor the results over time. This shift from simple task completion to autonomous execution marks a major turning point in AI's ability to function as an independent agent.

This ability to autonomously plan and execute tasks opens up a wide range of possibilities. In fields like business, finance, and customer service, Agentic AI can streamline operations, handle time-consuming processes, and significantly reduce the need for human oversight. For instance, in customer support, an AI could not only respond to inquiries but also track the progress of open issues, follow up on customer concerns, escalate matters when necessary, and even manage the scheduling of tasks. In the realm of software development, Gemini 2.0 can help developers write, test, and deploy code with minimal intervention, acting as an assistant that drives the development process forward.

For users, the autonomy of Agentic AI means more than just convenience. It means an AI that can understand their goals and needs on a deeper level and act on them without requiring constant supervision or instruction. Whether it's a personal assistant organizing a day's schedule, an AI-driven content generator that continuously produces new work, or a smart home system that adjusts to its user's habits, the power of Agentic AI lies in its ability to foresee the next steps and take action to create better outcomes with minimal input.

What makes Gemini 2.0's Agentic AI so powerful is the combination of its sophisticated reasoning abilities, multimodal capabilities, and access to vast amounts of data. Through these, it can understand a user's needs or a system's requirements, adapt to changes in real-time, and respond with an appropriate course of action. By continuously learning and refining its decision-making process, it can act more intelligently over time, becoming an even more efficient agent for its users.

Moreover, Gemini 2.0's Agentic AI can integrate seamlessly with other tools, systems, and platforms. It can coordinate with third-party applications, automate workflows, and even optimize processes across multiple domains. For example, a marketing team could use Gemini 2.0 to automatically generate and distribute campaign content, monitor user engagement, and adjust strategies in real time based on the feedback. It's not just about completing isolated tasks—it's about overseeing and executing end-to-end workflows.

In essence, Gemini 2.0's ability to function as an Agentic AI represents a transformative shift in the AI landscape. It's no longer a passive responder to user queries or requests; it's an active participant in task execution and decision-making. As this technology evolves, it promises to revolutionize industries, making businesses more agile, responsive, and capable of performing complex operations with less human oversight. Through its autonomous capabilities, Gemini 2.0 is pushing the envelope of what's possible, laying the foundation for a new era of intelligent, independent AI agents.

Project Astra and Project Mariner represent two groundbreaking initiatives that demonstrate

Gemini 2.0's potential to reshape how we interact with AI and the digital world. These projects aim to take the capabilities of Agentic AI to the next level, offering more personalized, efficient, and intelligent experiences for users.

Project Astra: A Universal Assistant with Memory and Context

Project Astra is designed to be a universal AI assistant that not only responds to queries but also remembers previous interactions, providing more personalized and contextually relevant responses. One of the standout features of Project Astra is its ability to recall past conversations, offering a more cohesive and continuous user experience. This long-term memory enables the assistant to provide tailored advice and solutions based on a user's history, preferences, and specific needs.

Unlike traditional AI assistants, which reset after each session and only remember the immediate context, Project Astra has the ability to learn and evolve based on its interactions over time. It can track ongoing projects, follow up on tasks, and even pick up where previous conversations left off, much like a human assistant would. Whether you're planning a long-term project or managing day-to-day tasks, Astra's memory ensures that it never loses track of your goals, allowing it to offer deeper insights and more relevant suggestions.

In addition to memory, Project Astra is integrated with powerful tools like **Google Search**, **Google Lens**, and **Google Maps**, which give it the capability to gather real-time information, visualize data, and navigate the physical world. For instance, if you're planning a trip, Astra can suggest the best routes using Maps, provide real-time traffic updates, and even recommend places to visit based on your preferences. With Google Lens, Astra can analyze images and provide detailed descriptions or contextual information, enabling a deeper understanding of visual content.

Moreover, the assistant's use of **Google Search** allows it to access a vast repository of online knowledge, ensuring that it can offer timely and accurate answers, no matter the topic. Whether you're seeking a product review, scientific research, or general advice, Project Astra's ability to contextualize the information it provides enhances the user experience. It can also prioritize responses based on a user's preferences, highlighting the most relevant and helpful information.

What makes Project Astra particularly powerful is its adaptability to different contexts. Whether it's helping you navigate a busy day, offering recommendations based on your habits, or providing personalized solutions for your work, the assistant's deep integration with various Google tools ensures that it can handle a broad range of tasks seamlessly. The assistant isn't just reactive; it anticipates your needs based on your past behaviors and ongoing projects, which can increase productivity and make everyday tasks much easier to manage.

Project Mariner: Navigating the Browser for Real-Time Task Completion

Project Mariner is a focused effort to expand Gemini 2.0's capabilities into the browser, enabling it to understand and interact with everything on a user's browser screen. This project is particularly useful for tasks that require real-time engagement with web content, such as filling out forms, navigating complex websites, or interacting with dynamic elements on web pages.

Where traditional AI assistants struggle to engage with content on the web directly, Project Mariner takes a more integrated approach, allowing Gemini 2.0 to perform actions within the browser environment itself. Whether it's a task as simple as submitting an online form or as complex as booking travel, Project Mariner enables seamless interaction between the user and the web interface.

For example, imagine you're booking a flight through a travel website. Rather than simply guiding you step-by-step, Project Mariner can analyze the website, understand your preferences (such as destination, dates, and budget), and automatically fill out forms or adjust options based on your input. If you decide to change your travel dates or explore different routes, Mariner can instantly adapt, making real-time adjustments without requiring you to manually input the information each time.

In addition to form-filling, Project Mariner excels at navigating complex websites with intricate structures. Websites that require multiple layers of interaction, such as online banking, shopping, or even job applications, can be difficult to navigate, especially when they feature dynamic content or need user input at various stages. Project Mariner can efficiently process these complex interactions, completing tasks in a fraction of the time it would take a user to manually enter data or explore multiple links.

This integration of Gemini 2.0 into the browsing experience also allows the AI to help users access hard-to-find information, such as detailed product specifications, reviews, or even hidden resources buried within websites. Mariner can navigate between tabs, open relevant links, and bring together information from different parts of the internet, offering a more cohesive experience. This isn't just about convenience; it's about saving time and energy by automating routine browsing tasks and making intelligent decisions in real-time.

Furthermore, Project Mariner also enables Gemini 2.0 to engage with dynamic content such as pop-ups, dropdown menus, and interactive features that change based on user input. It ensures a smooth, uninterrupted browsing experience, making sure that users can accomplish their tasks without frustration or delays.

Together, Project Astra and Project Mariner highlight the full range of Gemini 2.0's versatility. Whether you're interacting with a conversational AI that remembers your preferences and uses external tools to deliver highly personalized content, or you're navigating the complexities of the web with an AI that completes tasks and fills in forms on your behalf, these projects demonstrate how Gemini 2.0 is more than just a tool. It is evolving into a fully capable, intelligent agent capable of performing complex tasks in dynamic environments—transforming the way we interact with AI in our everyday lives.

Jewel – An AI-powered Coding Agent

Jewel is a powerful addition to the AI landscape, specifically designed for developers who need an intelligent assistant to streamline their coding tasks. Unlike traditional coding tools, Jewel acts as an AI-powered coding agent that seamlessly integrates with platforms like **GitHub**, enabling developers to optimize their workflows and enhance productivity. It is not just another code generation tool; Jewel is designed to understand complex coding tasks and execute them with precision, helping developers focus on building more complex features rather than getting bogged down by repetitive tasks.

At its core, Jewel functions as a **code assistant** that can analyze existing codebases, suggest improvements, debug errors, and even generate new code based on the user's requirements. It leverages the power of **Gemini 2.0's multimodal capabilities** to interpret and understand both code syntax and the intent behind the code. Jewel is able to process complex queries and provide solutions that are contextually relevant and efficient, making it an indispensable tool for developers across different stages of the software development lifecycle.

One of the standout features of Jewel is its **integration with GitHub**, a platform widely used by developers to host and collaborate on code. This integration allows Jewel to interact directly with the code repositories stored on GitHub, pulling in data, understanding the existing code structure, and even suggesting enhancements or alternative approaches to solving problems. Whether it's fixing bugs, optimizing algorithms, or refactoring code to improve performance, Jewel helps developers stay on top of their coding tasks by offering intelligent, context-aware suggestions.

Jewel's ability to **execute coding tasks autonomously** also sets it apart from other AI tools. For instance, it can handle routine tasks such as code formatting, variable naming conventions, or even generating boilerplate code. This automation not only saves developers valuable time but also ensures that the generated code adheres to best practices and is free of common errors. Developers no longer need to spend hours on mundane tasks; Jewel takes care of the heavy lifting, allowing them to focus on the creative and innovative aspects of their projects.

Moreover, Jewel doesn't just suggest improvements—it also learns from the developer's preferences and habits. Over time, it becomes more adept at understanding the developer's coding style and the specific needs of each project. This dynamic learning capability allows Jewel to offer more personalized and relevant advice, making it an increasingly valuable tool as developers continue to use it.

With its sophisticated natural language processing abilities, Jewel can understand **complex code-related queries** and convert them into executable tasks. Developers can simply describe what they need in plain language, and Jewel will generate the appropriate code, making it accessible even for those who may not be experts in every programming language. This feature can help reduce the learning curve for new developers while enhancing the capabilities of seasoned professionals.

Jewel's integration with **GitHub Actions** further enhances its utility. By automating parts of the development workflow, such as continuous integration (CI) and continuous deployment (CD), Jewel helps streamline the development pipeline. It can detect issues in the code before they reach the production environment, ensuring that only high-quality, bug-free code makes it through the development cycle.

As the demand for efficient, high-quality software development continues to grow, tools like Jewel become invaluable in the developer toolkit. By combining the power of **AI with existing development platforms like GitHub**, Jewel offers a transformative approach to coding, enabling faster, smarter, and more effective development processes. It's an AI assistant that doesn't just understand code—it becomes an integral part of the development journey, helping developers write, debug, and optimize code with minimal friction and maximum efficiency.

Chapter 5: AI in the Real World – Transforming Industries

Gemini 2.0 is poised to have a transformative impact on a variety of industries, with the gaming sector standing out as one of its most exciting applications. The AI's advanced capabilities are opening new avenues for enhancing gameplay and reshaping the gaming experience, both for developers and players. By leveraging its ability to process multimodal data and execute real-time reasoning, Gemini 2.0 is not only changing how games are designed, but also how they are experienced by players.

In the gaming world, one of the most thrilling prospects of Gemini 2.0 is the ability to **analyze real-time game data** and offer strategic advice tailored to each player's actions. For players, this means having an AI assistant capable of providing personalized insights, helping them make better decisions in complex game scenarios. Whether it's offering tips for improving gameplay or suggesting strategies in competitive multiplayer games, Gemini 2.0 is redefining the role of AI in gaming.

But its influence doesn't stop there. **AI agents** powered by Gemini 2.0 can also enhance team dynamics by suggesting optimal team compositions based on player strengths, roles, and playstyles. This is particularly relevant in team-based games, where synergy between players is key to success. By analyzing individual player behavior and real-time data from the game environment, Gemini can help form teams that complement each other, offering a more balanced and strategic approach to gameplay. This dynamic team-building feature goes beyond simple matchmaking, taking into account a wealth of factors that influence team performance, such as player communication, in-game actions, and overall strategy.

Moreover, **game development itself** benefits immensely from Gemini 2.0. With its ability to generate complex narratives, create dynamic game environments, and offer real-time suggestions, game designers can lean on AI to enhance creative processes. Whether generating dialogue options for interactive storytelling or adjusting the difficulty curve based on player performance, Gemini 2.0 provides tools that can vastly improve the gaming experience, making it more immersive and engaging.

Shifting from the virtual world of gaming to the physical world, Gemini 2.0 is also making waves in fields like robotics, manufacturing, and healthcare. The underlying technology behind Gemini's multimodal capabilities—its ability to understand and interpret diverse forms of input, including visual and spatial data—makes it a powerful tool in industries that rely on complex physical and environmental interactions.

In **robotics**, for instance, Gemini 2.0 can help machines navigate and interact with physical environments with greater accuracy and autonomy. By processing real-time visual and sensor data, Gemini's advanced spatial reasoning allows robots to understand their surroundings and adjust their actions accordingly. This could have profound implications for industries such as warehousing, logistics, and assembly lines, where robots can perform tasks like sorting, organizing, or assembling products with minimal human intervention. The integration of **AI with physical environments** not only increases efficiency but also reduces the risks associated with human labor in hazardous settings, such as factories or construction sites.

In **manufacturing**, Gemini 2.0 could improve production line efficiency by predicting failures before they happen, optimizing workflows, and ensuring that quality control measures are more accurate and faster. AI agents can help monitor machines, track performance, and suggest adjustments to improve productivity while reducing downtime and operational costs. By incorporating machine learning models that learn and adapt to changing conditions, Gemini can enhance the agility and flexibility of manufacturing operations.

One of the most exciting applications of Gemini 2.0 is its potential in **healthcare**. The AI's ability to analyze vast amounts of medical data and interact with physical environments makes it a powerful tool for patient care. For example, robots powered by Gemini 2.0 could assist with surgeries, offering real-time analysis of patient data and providing guidance to doctors during operations. In healthcare settings, where precision and speed are critical, this kind of AI assistance could enhance outcomes and improve the accuracy of diagnoses. Similarly, in areas like **remote patient monitoring**, Gemini's AI can help track patient vitals, detect anomalies, and provide feedback to healthcare professionals in real-time, all while learning from the data to offer better recommendations in the future.

As we look to the future, it's clear that the full potential of Gemini 2.0 in robotics, manufacturing, and healthcare is just beginning to be realized. Its ability to process multimodal inputs, reason spatially, and execute real-time actions in physical environments will likely revolutionize these industries. What we are witnessing now is just the tip of the iceberg, with AI like Gemini 2.0 set to reshape how we work, interact, and even live. From improving manufacturing workflows to assisting in life-saving medical procedures, the possibilities for Gemini 2.0 are vast and continue to unfold at an exciting pace.

The future of logistics is set to be significantly reshaped by the capabilities of Gemini 2.0, unlocking unprecedented levels of automation, efficiency, and safety within the industry. As global trade and the movement of goods continue to grow more complex, companies are turning to advanced AI technologies like Gemini 2.0 to streamline operations and meet the demands of an increasingly dynamic marketplace.

One of the key advantages Gemini 2.0 brings to **logistics** is its ability to analyze and process multimodal data in real time. This capability enables AI-powered systems to **optimize supply chains**, track shipments, and manage inventory with remarkable accuracy. By integrating real-time data from a variety of sources—such as sensors on trucks, GPS data, and weather patterns—Gemini can provide logistics managers with a comprehensive view of operations. This holistic perspective allows for more **precise routing**, reducing fuel consumption and travel times while ensuring that goods arrive on schedule.

Beyond simple tracking, Gemini 2.0's ability to process multimodal inputs allows it to make proactive decisions about operations. For example, if an unforeseen weather event threatens to delay shipments, Gemini can automatically reroute vehicles, adjust schedules, and even communicate with customers to manage expectations. This level of **autonomous decision-making** not only enhances efficiency but also boosts the overall reliability of logistics systems.

In **transportation management**, Gemini 2.0 can also contribute to **safety** improvements. AI-driven systems can monitor driver behavior, road conditions, and vehicle performance in real-time, identifying potential safety risks before they become critical issues. By analyzing these inputs, Gemini can alert drivers to potential hazards, suggest safer routes, or even make adjustments to vehicle systems to prevent accidents. For instance, in **self-driving trucks**, Gemini 2.0's AI can guide the vehicle through complex environments, detect obstacles, and adapt to changing conditions, minimizing human error and reducing the likelihood of accidents.

Additionally, **warehousing and distribution** centers will also benefit from the AI's multimodal capabilities. Gemini 2.0 can help optimize the storage and retrieval of goods by autonomously managing inventory, tracking product movements, and forecasting demand. In a fast-paced distribution center, this can dramatically reduce wait times and ensure products are available when needed, thereby improving **turnaround times** and reducing overhead costs.

As **autonomous vehicles** and **drones** become increasingly common in logistics, Gemini 2.0's ability to interpret complex data feeds from these devices will be critical in coordinating their activities. Whether it's overseeing fleets of delivery drones or coordinating autonomous trucks on highways, Gemini's advanced spatial reasoning and real-time decision-making can help ensure that all operations run smoothly and safely.

Looking further into the future, Gemini 2.0's potential to **optimize entire supply chains** could fundamentally change how goods are produced, transported, and delivered globally. By continuously learning from vast amounts of operational data, the AI will be able to predict and prevent disruptions, streamline workflows, and help businesses respond faster to changing market conditions. Whether it's enhancing last-mile delivery systems, improving inventory management, or optimizing transportation routes, Gemini 2.0 will play a crucial role in building smarter, more resilient logistics networks.

As the logistics industry increasingly relies on automation and AI to meet the demands of a growing global economy, Gemini 2.0 is positioned to be at the forefront of this revolution. Its ability to not only understand and process multimodal data but also to act on it autonomously will be key to unlocking new levels of operational excellence, safety, and customer satisfaction. The future of logistics is brighter and more efficient with AI systems like Gemini 2.0 leading the way.

Chapter 6: Ethical Considerations and Safety Measures

Google has always been committed to ensuring that the technologies it develops are safe, ethical, and responsibly deployed. With the advent of **Gemini 2.0**, this commitment has only become more pronounced. As artificial intelligence continues to play a greater role in daily life, the risks associated with its misuse or unintended consequences become increasingly significant. Therefore, Google has taken multiple proactive steps to ensure that Gemini 2.0 is not only powerful but also used in a way that is secure, responsible, and in line with ethical standards.

One of the key pillars of Google's approach to AI safety is its **extensive testing** processes. Before launching Gemini 2.0 into the world, Google subjected the AI to a series of rigorous tests, simulating real-world applications across a variety of use cases. These tests aim to identify any potential vulnerabilities, biases, or unintended behaviors the AI might exhibit. Google's dedicated teams run these tests with the goal of catching issues before they affect real users, ensuring that the AI behaves in a predictable and reliable manner.

Moreover, Google has formed a **Safety and Responsibility Committee**, which is responsible for overseeing the ethical deployment of AI technologies like Gemini 2.0. This committee is composed of experts from various fields, including **AI ethics**, **law**, **privacy**, and **sociology**. Their role is to provide guidance on the development and implementation of policies that ensure AI technologies are used responsibly and do not inadvertently cause harm to individuals or society. This group reviews potential risks associated with AI models and provides actionable recommendations to mitigate these risks. Whether it's ensuring that **AI outputs** are free from **biases** or addressing **privacy concerns**, this committee plays a crucial role in steering the direction of Google's AI development.

Another key aspect of Google's approach to safety is its **collaboration with external organizations** and stakeholders. Google understands that the future of AI is a shared responsibility, and as such, it works alongside academic institutions, policymakers, and non-profits to promote the development of safe AI practices. This broad approach ensures that a diversity of perspectives is considered when shaping the future of AI.

While the safety of AI models is paramount, **privacy** remains one of the most important concerns for users. **Gemini 2.0** is designed with a host of **privacy safeguards** to ensure that user data is protected and that individuals retain full control over their interactions with the system. One of the key features is **user control over memory**. Unlike earlier versions of AI, Gemini 2.0 allows users to manage what the AI remembers, offering an unprecedented level of transparency and control.

If users are uncomfortable with the AI storing any information, they can choose to delete specific conversations or even opt for a setting where the AI doesn't retain any memory at all after each interaction. This allows for a more **dynamic and customizable** experience, where users can feel confident that their personal data is treated with respect.

Google has also integrated several **security features** into Gemini 2.0 to ensure that data is kept safe. **End-to-end encryption** is utilized to protect sensitive conversations, ensuring that data cannot be accessed or intercepted during transmission. Additionally, Google provides users with a clear and **transparent privacy policy**, detailing how data is collected, stored, and used by the system. This transparency allows users to make informed decisions about how they interact with Gemini 2.0.

For businesses and developers integrating Gemini 2.0 into their platforms or workflows, Google also offers tools to help them comply with **data protection regulations** such as **GDPR**. These tools allow organizations to **anonymize** user data and ensure that they are only storing information that is necessary for the intended purpose, further enhancing privacy protections.

Another important feature in the realm of privacy is that users can **view and manage their data** via a simple interface. This feature allows individuals to see what information has been retained by the AI, providing a clear view of any data that has been stored. With a few clicks, users can delete conversations, disable the AI's memory, or even request that all data be wiped from the system, empowering individuals to protect their personal information. These features reinforce Google's commitment to **user autonomy**, giving users complete control over what they share with Gemini 2.0.

The company's approach to privacy doesn't just stop with the user-facing features. Behind the scenes, Google adheres to **strict data governance protocols** to ensure that any data collected is handled responsibly and securely. The company also ensures that its AI systems are regularly audited by both internal teams and external organizations to confirm that they remain compliant with **international data privacy standards**.

As AI continues to evolve, **Google** recognizes that user trust is critical to the widespread adoption of these technologies. By taking a proactive and transparent approach to safety and privacy, the company is helping to ensure that **Gemini 2.0** remains a responsible and trusted tool for users across the globe. Whether it's providing clear control over data, ensuring the security of communications, or actively working to mitigate potential risks, Google's commitment to responsible AI development is integral to the success of **Gemini 2.0** in the years to come.

As artificial intelligence continues to evolve and become more deeply integrated into our daily lives, ensuring that AI is used ethically and safely is more important than ever. Google has placed a strong emphasis on both **security** and **ethical considerations** when developing **Gemini 2.0**, with the goal of creating a tool that not only serves its users effectively but also prevents potential harm and misuse. This is why Gemini 2.0 is designed with advanced mechanisms to **recognize and avoid malicious instructions** and promote **fairness** and **transparency** in its interactions.

One of the most crucial aspects of preventing misuse lies in Gemini 2.0's ability to **detect malicious behavior.** In today's digital world, **phishing attacks, scams,** and other forms of online deception are common, and AI systems can often be a target or even an unwitting accomplice in such schemes. Gemini 2.0, however, has been designed with sophisticated safeguards that allow it to **recognize harmful requests** and **avoid**

participating in malicious activities. For example, if a user tries to instruct Gemini 2.0 to generate a phishing email or engage in other unethical behaviors, the AI will be able to detect the malicious intent behind the request. Rather than complying, Gemini 2.0 will either **refuse to act** on the request or **warn the user** about the potential dangers of their actions.

This built-in safeguard acts as a layer of protection for users, preventing them from inadvertently becoming involved in cybercrime or other harmful activities. Moreover, it reduces the risk of the AI itself being manipulated into supporting malicious schemes, something that is critical as AI systems become more widely used in various contexts, from customer support to content generation. Google has ensured that Gemini 2.0's **malicious instruction detection system** is not only reactive but also continuously improved through **machine learning** to stay ahead of evolving threats, thereby ensuring a **robust defense** against potential misuse.

But AI security is just one side of the coin. The other half lies in ensuring that the AI operates in a **fair**, **transparent**, and **ethical** manner. **Ethics in AI** is a growing concern, particularly as AI systems become more powerful and begin to influence decisions that can affect people's lives in profound ways. Google's approach to ethical AI revolves around a set of core principles that ensure Gemini 2.0 works in a way that is **just, inclusive**, and **accountable**.

At the heart of this ethical approach is the idea that AI should be designed and used in a way that **respects the dignity** and **rights** of all individuals. This is why Gemini 2.0 has been engineered to operate transparently, ensuring that users are fully aware of how it works, what data it collects, and how that data is used. **Transparency** is a fundamental aspect of ethical AI, as it allows users to make informed decisions about how they interact with AI systems. For example, users can easily access information about how Gemini 2.0 processes their requests, what algorithms drive its responses, and how their data is handled. This transparency is crucial for building trust and confidence in the AI, as it ensures that there are no **hidden agendas** or **unseen biases** at play.

Fairness is another key pillar in the ethical framework for Gemini 2.0. The development of the AI takes into account the potential for **bias** in the data it is trained on. Since AI models learn patterns from data, there is always the risk that the data used to train these models might be skewed, reflecting biases that could perpetuate discrimination or inequality. Google has taken significant steps to **mitigate bias** in Gemini 2.0 by ensuring that the training datasets are diverse and representative of a wide range of perspectives. Additionally, the company regularly audits the AI's performance to identify any unintended biases that may emerge, working to continuously improve the model's fairness.

In addition to fairness and transparency, **accountability** is an important consideration in the development of Gemini 2.0. Google is committed to ensuring that the AI behaves responsibly and does not cause harm to individuals or society. This means that there are **clear accountability structures** in place, with both internal and external mechanisms to monitor the AI's actions and ensure compliance with ethical standards. For example, Google's **Safety and Responsibility Committee** oversees the ethical implications of the AI's deployment and usage, ensuring that it aligns with the company's values and societal expectations.

Gemini 2.0's **ethical design** is also focused on **user empowerment**. The AI is created with the understanding that it should serve people, not replace or dominate them. As a result, Gemini 2.0 provides users with control over their interactions, allowing them to manage their data, customize their experience, and decide what the AI retains. This user-centric design ensures that the AI is not only useful but also **respectful** of individual preferences and privacy.

By combining robust **security features** with an unwavering commitment to **ethical principles**, Gemini 2.0 sets a new standard for responsible AI deployment. Google's focus on **preventing misuse**, **avoiding biases**, and ensuring **transparency** and **fairness** makes Gemini 2.0 not just a powerful tool, but also one that users can trust and feel safe using. With these safeguards in place, Gemini 2.0 is poised to shape the future of AI, helping to guide it toward a path that benefits everyone.

Chapter 7: The Competitive Landscape

In the rapidly evolving world of artificial intelligence, **Gemini 2.0** is entering a competitive arena, vying for dominance against some of the most well-known AI models on the market today. Leading the charge are companies like **OpenAI**, **Microsoft**, and **Anthropic**, each of which has its own powerful AI offerings that are changing the way we think about AI's role in both business and daily life. While all of these models share certain capabilities, Gemini 2.0 brings a distinct set of advantages to the table that could make it the preferred choice for many users and developers alike.

One of the most **striking strengths** of **Gemini 2.0** lies in its **multimodal capabilities**. While models like **OpenAI's GPT-4** are incredibly proficient at text-based tasks, Gemini 2.0 goes several steps beyond by enabling the creation of **multimodal outputs**. This means that users can not only interact with Gemini 2.0 via **text** but can also generate **images**, **audio**, and **videos**—all seamlessly integrated within the same interaction. This type of **multimodal functionality** gives Gemini 2.0 a distinct edge in scenarios that demand more than just a simple text response. Whether it's generating a **video script with visuals**, creating **audiovisual customer support** solutions, or offering **dynamic content for marketing** purposes, Gemini 2.0 can handle tasks that go beyond traditional text-based AI.

Moreover, the integration of **Gemini 2.0** into Google's **vast ecosystem** is another area where it outshines the competition. While OpenAI's models are deeply integrated with **Microsoft's products** like **Word** and **Excel** through **Copilot**, and **Anthropic's Claude** is making strides in building out conversational AI, Gemini 2.0 benefits from its **tight-knit connection** with other Google tools, such as **Google Search**, **Google Maps**, **Google Lens**, and **Google Workspace**. This powerful integration allows **Gemini 2.0** to deliver an experience that is not only **convenient** but also **highly productive**, as users can access Google's best-in-class products while interacting with the AI. For example, whether you're using Gemini 2.0 for **content generation**, **research**, or even **multilingual customer support**, its seamless connection with **Google Search** means it can leverage vast amounts of data in real time to deliver contextually relevant and highly accurate results.

However, despite these strengths, Gemini 2.0 faces significant challenges when it comes to **brand recognition** and **user adoption**. OpenAI's **GPT-4** and **Anthropic's Claude** are already **well-established players** in the AI market, and both have garnered significant **user loyalty** and **brand trust**. OpenAI, in particular, has the advantage of having been among the first to develop and release GPT-based models, which have made huge waves in both business and popular culture. Microsoft's partnership with OpenAI only amplifies this visibility, as the AI is integrated into widely used tools like **Microsoft 365** and **GitHub Copilot**, making it difficult for competitors like **Gemini 2.0** to quickly catch up.

Similarly, **Anthropic's Claude**, which has gained traction due to its **focus on safety and transparency** in AI interactions, also has a strong position in the market, particularly among users who prioritize ethical AI design. These established players have not only created **advanced AI models** but have also built strong communities and support ecosystems around their tools, contributing to **higher user adoption rates** and **brand loyalty**.

For **Gemini 2.0** to stand out in such a competitive field, it needs to overcome these **brand awareness challenges**. Google, despite its dominance in the tech world, is still relatively new to the AI field in comparison to OpenAI and Anthropic. In order to gain a larger user base, **Google will need to emphasize Gemini 2.0's unique advantages** and promote its multimodal capabilities and deep integration into the broader Google ecosystem. The company could leverage its existing user base from products like **YouTube**, **Google Search**, and **Google Assistant** to introduce Gemini 2.0 to a wider audience. By demonstrating how Gemini 2.0 can enhance productivity, creativity, and everyday tasks, Google can build awareness and attract new users.

Additionally, **education** and **developer engagement** will play key roles in Gemini 2.0's success. Google can focus on empowering **developers** with tools like the **Gemini API**, **Google AI Studio**, and **Vertex AI**, which enable them to easily experiment with and integrate Gemini 2.0 into their own applications and services. By encouraging developers to build on top of Gemini 2.0 and create compelling use cases, Google can increase the **adoption rate** among businesses, startups, and tech enthusiasts, driving **word-of-mouth awareness** and improving brand recognition.

Another potential strategy for Google is **collaborating with other industries** to integrate Gemini 2.0 into **enterprise-level solutions**. For instance, companies in sectors like **healthcare**, **education**, **entertainment**, and **e-commerce** could benefit greatly from Gemini 2.0's **multimodal capabilities** and the ability to provide **personalized, context-aware** AI solutions. By demonstrating these practical applications, Google can show how Gemini 2.0 can help businesses across industries solve complex challenges while offering a competitive advantage.

In the long run, **user adoption** will depend on how effectively Google communicates the **value proposition** of Gemini 2.0 to a wide range of potential users. Whether it's through increased visibility, educational initiatives, or strategic partnerships, the goal will be to make Gemini 2.0 not just another AI tool, but a **game-changer** that drives the future of **artificial intelligence** for both **developers** and **end-users**. By highlighting its strengths in **multimodal functionality, speed**, and **integration with Google's suite of products**, Gemini 2.0 can carve out a niche for itself in a crowded market and ultimately become a leader in the next wave of AI innovation.

The future of artificial intelligence stands at the edge of a profound transformation, and as AI continues to evolve at a rapid pace, the role of models like **Gemini 2.0** will be integral to shaping what's next in the world of technology. AI is no longer confined to simple task automation or answering queries—it is on the verge of revolutionizing industries, reimagining how humans interact with technology, and offering unprecedented solutions to complex challenges.

As we look ahead, the future of AI can be broken down into several key areas: **enhanced human-AI collaboration, ethical and responsible AI development, AI-driven automation** across industries, and the seamless integration of AI into daily life. In each of these areas, Gemini 2.0 is well-positioned to play a pivotal role, not just as a tool, but as a collaborator in the true sense of the word.

The future of **human-AI collaboration** is one of the most exciting frontiers. As AI models become more sophisticated, they will no longer just be **tools** that execute commands—they will act as **partners** capable of co-creating, problem-solving, and driving innovation alongside human expertise. This collaboration will be transformative, enabling people to achieve things that would be impossible on their own. **Gemini 2.0**, with its multimodal capabilities, represents a major leap in this direction. The AI's ability to process not just text, but also **images**, **videos**, and **audio**, allows it to become a true creative collaborator. For artists, filmmakers, content creators, and developers, Gemini 2.0 will be able to **generate content across multiple formats**, offering more than just written responses. Whether it's generating detailed **visual narratives**, assisting in **video creation**, or providing **real-time multilingual translations**, Gemini 2.0's multifaceted approach will allow people to create and innovate in ways they never could before.

This shift from AI being a **simple assistant** to a **creative partner** is just one aspect of a broader trend toward **AI-powered automation** across all industries. **Gemini 2.0's capabilities** extend beyond entertainment and content creation—its **spatial reasoning**, **advanced decision-making abilities**, and **autonomous task execution** are set to disrupt sectors such as **healthcare, education, logistics**, and **finance**. In healthcare, for example, Gemini 2.0 could support doctors by **analyzing patient data**, **identifying patterns**, and suggesting treatment plans. In logistics, it could **optimize delivery routes** or **predict supply chain issues**. The more versatile AI becomes, the more integrated it will be in industries that rely on efficiency and data-driven decision-making.

One of the most exciting implications of AI's future is its potential to **enhance productivity** and streamline workflows in ways that are unimaginable today. With the emergence of **Agentic AI**, such as the features found in **Gemini 2.0**, AI will evolve from a passive assistant to an active, **goal-oriented agent** capable of planning and executing tasks independently. This could revolutionize everything from office workflows to customer service, with AI taking on an increasingly proactive role. Gemini 2.0's ability to **anticipate needs**, **make decisions**, and **integrate with other services** will be invaluable in creating smarter workflows and automating routine tasks. This will free humans to focus on more **strategic**, **creative**, and **value-added tasks**, which could lead to an **explosion of innovation** in the years to come.

But as AI grows more powerful, it also raises questions about **responsibility** and **ethics**. One of the key challenges will be ensuring that AI evolves in a way that is both **transparent** and **fair**. Here, **Google's commitment** to **ethics in AI** and its focus on **safe AI development** are critical. With models like Gemini 2.0, there is a strong emphasis on ensuring that the AI does not perpetuate **bias**, and that its use is **secure**, **private**, and **accountable**. This type of ethical responsibility will be crucial as AI begins to take on more critical tasks, from medical diagnosis to autonomous driving. **Google's AI Safety** protocols and **privacy safeguards** will be essential to ensuring that the advancements Gemini 2.0 represents do not come at the cost of user privacy, security, or fairness.

Looking further into the future, **Gemini 2.0** could play a major role in the **evolution of AI consciousness**, where AI could potentially become aware of its capabilities, limits, and roles in society. While this is still in the realm of speculation, AI's potential to **reflect on its actions**, **learn autonomously**, and **adapt to its**

environment could open up new avenues for human-AI collaboration. Over time, we may see AI models like Gemini 2.0 move beyond task completion, evolving into more **sophisticated forms of artificial intelligence** capable of **emotional intelligence, self-improvement**, and perhaps even **personalized AI interactions** tailored to individual users' preferences and needs.

Gemini 2.0 also holds the potential to push the boundaries of AI's role in education and research. With its deep learning capabilities and the ability to process complex data across multiple modalities, it could become a key tool for **researchers, students**, and **educators** in every field. Imagine an AI capable of not only explaining concepts but also **visualizing** them in real-time, generating 3D models, creating simulations, and offering personalized learning experiences tailored to the **individual's** pace and style. It could serve as a tutor that adapts to the learner's needs, providing a deeper, more interactive education system than what is possible today.

In terms of **growth**, **Gemini 2.0** is just at the beginning of its journey. With the rapid pace of technological advancements in AI, particularly in areas like **neural networks**, **machine learning**, and **cloud computing**, we can expect to see Gemini 2.0 evolve quickly. As more users engage with it, the model will continuously improve through **machine learning** and **user feedback**, becoming increasingly powerful and versatile over time.

In conclusion, the future of **Gemini 2.0** is bright, filled with exciting possibilities. It stands as a powerful tool in the growing realm of **multimodal AI**, poised to shape how we work, create, and interact with machines. As AI continues to grow more intelligent, efficient, and autonomous, it will have an even more profound impact on the way humans live and work—leading to a future where AI and human collaboration can unlock opportunities that were once thought impossible. Whether it's enhancing creativity, revolutionizing industries, or ensuring ethical standards, **Gemini 2.0** will play an instrumental role in defining the next era of artificial intelligence.

Chapter 8: What's Next for Gemini 2.0 and AI?

As we look toward the future of Gemini 2.0, there is an unmistakable excitement surrounding the potential for even more groundbreaking advancements. **Gemini 2.0** has already demonstrated its capabilities, but as with all AI models, it is constantly evolving, and new features and improvements are on the horizon. This ongoing development process is fueled by the feedback from both **developers** and **users**, as well as the continuous progress in AI research and technology.

Upcoming features and updates for Gemini will likely address both the technical and user experience aspects of the platform, pushing its boundaries even further. One area that is poised for significant improvement is **Agentic AI**. While **Gemini 2.0** already demonstrates the ability to plan and execute tasks autonomously, future versions could introduce even **more sophisticated autonomous agents**, capable of handling increasingly complex scenarios. These agents might be able to **anticipate user needs** with greater accuracy, **optimize workflows** across multiple domains, and **integrate seamlessly** with even more third-party platforms and tools, making them not only highly efficient but also **contextually aware** of users' environments.

Another expected improvement lies in the **multimodal capabilities** of Gemini. Currently, Gemini can generate multimodal outputs such as images, videos, and audio, alongside text. However, future updates may enhance its ability to **generate even more dynamic and interactive content**, incorporating virtual and augmented reality (VR/AR) into its creative suite. Imagine Gemini being able to not only generate text-based content or visuals but also interactive AR or VR simulations that immerse users in digital environments. These advancements could transform industries such as gaming, education, and entertainment, where the lines between **the physical** and **digital worlds** continue to blur.

In terms of **speed and efficiency**, we can anticipate even faster **processing times** as well as improvements in **accuracy** and **scalability**. As AI models grow in complexity, ensuring that Gemini remains **responsive and agile** while handling large amounts of data will be crucial. One avenue for achieving these improvements could lie in the continuous optimization of Google's **Trillium TPUs**, which provide the hardware foundation for Gemini 2.0's powerful capabilities. In future versions, these processors may become even more advanced, enabling **faster real-time responses** and **better handling of large-scale tasks** across industries such as healthcare, logistics, and manufacturing.

As for **AI ethics** and **privacy**, we can expect **continued innovation** in how **Gemini 2.0** handles sensitive data. With privacy being an increasing concern for users, future updates will likely emphasize even **more robust privacy safeguards** and **user control** over data management. For instance, improvements could include **enhanced mechanisms** for users to control what the AI remembers, delete interactions, and ensure that AI-generated content adheres to ethical guidelines. Furthermore, **AI transparency** will become a key focus, with Google providing more **explainability features** so users can better understand how decisions are made, especially in high-stakes industries such as finance or healthcare.

Looking beyond Gemini 2.0, the **pace of AI advancements** is only accelerating. We are on the cusp of seeing a future where **Agentic AI** goes beyond task automation and begins to engage in more complex forms of human-AI collaboration. As the technology matures, we might see future versions of **Gemini** capable of understanding and making decisions with even greater depth, fostering an entirely new way for humans and AI to co-exist.

This could involve AI agents that act with **increased autonomy**—whether that's helping to run businesses, providing real-time feedback for education, or creating personalized content that evolves with a user's changing needs.

Moreover, **Gemini 2.0** could integrate even more seamlessly across industries, creating solutions that span beyond the boundaries of individual sectors. The world is shifting toward an interconnected ecosystem where industries rely on cross-platform technologies, and **AI-powered integration** is central to this transformation. For example, Gemini could expand its role in **smart cities**, supporting everything from **automated traffic management systems** to **energy-efficient urban planning**, taking advantage of the interconnected data streams within these complex systems.

The **evolution of Agentic AI** could also lead to deeper integration into more aspects of our daily lives, with AI taking on more sophisticated roles in

healthcare (from diagnosis to personalized treatment plans), **manufacturing** (automating entire production lines), and even **artificial creativity** (evolving alongside human artists to create immersive, dynamic storytelling experiences).

Ultimately, Gemini 2.0 is part of a much larger trend toward **autonomous AI systems** that are no longer just tools, but rather collaborators capable of transforming entire industries. The potential for these advancements is limitless, and as we continue to innovate, the boundary between what is possible with AI and what is beyond reach will continue to shift, moving us closer to an era of **AI-driven creativity, intelligence,** and **autonomy**.

As we await the next generation of **Gemini 2.0** and the unfolding landscape of AI, one thing is certain: the journey has only just begun. We can expect to see continued progress in AI's capabilities to not only **understand and create** but to **evolve**

autonomously, shaping a future where **human and machine** can collaborate more effectively than ever before. With every new iteration, Gemini will redefine the possibilities of what AI can do, and the way we interact with it, making it an exciting time for both the industry and its users.

As we reflect on the transformative power of AI and the groundbreaking advancements embodied in **Gemini 2.0**, it becomes clear that we are standing at the threshold of a profound shift. For everyday users and society as a whole, the implications of AI's rapid evolution will be far-reaching, touching every aspect of our lives in ways we may not yet fully comprehend. The changes we are witnessing today are merely the beginning of what promises to be a revolution in how we live, work, and interact with technology.

In the realm of **work**, AI's integration into industries like finance, healthcare, manufacturing, and logistics is already making a significant impact. As AI systems like Gemini 2.0 become more adept at handling complex tasks and making autonomous decisions, **human workers** will be freed from repetitive, mundane tasks, enabling them to focus on more **creative, strategic, and value-added roles**. AI will serve as an **intelligent assistant**, helping to optimize workflows, analyze vast amounts of data, and even predict trends and outcomes with far greater accuracy. This shift will require workers to adapt, embracing the opportunity to work alongside machines rather than view them as competitors. The future of work will likely see a blending of **human ingenuity** with **machine precision**, creating more **collaborative environments** where both can thrive.

On a societal level, AI's influence will extend into **education**, reshaping how we learn and access knowledge. Tools like Gemini 2.0's **multimodal capabilities** could revolutionize the classroom, providing personalized learning experiences that cater to the unique needs of each student. Imagine a world where an AI-driven tutor can assess a student's learning style, identify their strengths and weaknesses, and deliver customized lessons in real-time, helping them to grasp complex concepts with ease. This **personalized education** model could make learning more accessible, ensuring that everyone, regardless of background or ability, has the opportunity to succeed. AI could also democratize education by breaking down geographical and financial barriers, giving access to world-class resources, expertise, and mentorship to anyone with an internet connection.

Beyond work and education, AI will have an undeniable influence on how we **interact with technology** in our everyday lives. The introduction of **Agentic AI**—systems that not only respond to our commands but anticipate our needs and execute tasks autonomously—will dramatically alter the landscape of personal devices. Your smartphone, your home assistant, your car—each of these will evolve into far more capable and intelligent systems that can **adapt** to your preferences and **predict** your actions. Imagine a device that not only answers your questions but also **organizes your day, makes recommendations**, and even **helps you set and achieve goals**. The AI embedded in these devices will learn from your behavior, becoming more personalized and **context-aware** as time goes on. These advancements will blur the lines between what we consider **human** and **machine**, as technology becomes more intuitive and seamlessly integrated into our daily routines.

However, as we embrace these innovations, it's important to recognize that the integration of AI into society comes with its own set of challenges. **Ethical concerns** regarding **privacy, security**, and **bias** will need to be addressed. While AI can unlock immense potential, we must ensure that it is used responsibly and for the benefit of all. Issues like data privacy, algorithmic fairness, and transparency will continue to be central to the discourse around AI development. Google's efforts with **Gemini 2.0** to **ensure privacy safeguards**, prevent misuse, and focus on **ethical AI** are steps in the right direction, but the global conversation on these topics must continue.

One of the most profound implications of **AI advancements** is their potential to **reshape industries** and even **entire economies**. As machines become more capable of handling complex tasks, industries that were once dependent on manual labor or human expertise will evolve. **Healthcare**, for instance, will see AI assisting with diagnostics, treatment planning, and even surgery, improving outcomes and accessibility. In **manufacturing**, AI-driven robots will streamline production lines, enhance precision, and reduce costs. Meanwhile, **logistics and transportation** will be revolutionized by autonomous vehicles, drones, and AI systems capable of managing and optimizing supply chains with unprecedented speed and efficiency. As these industries transform, new job opportunities will emerge, focused on **AI development**, **robotics**, and **AI ethics**, while others will evolve to leverage the **synergy between humans and machines**.

Perhaps the most important consideration is the **societal impact** of these changes. The introduction of **autonomous AI** systems has the potential to alter the fabric of society itself, affecting everything from the **job market** to our **personal relationships** with technology. As AI continues to become more integrated into our daily lives, the boundaries between the **digital** and **physical worlds** will continue to blur. We will be interacting with AI in ways that feel almost **second nature**, whether it's through a voice assistant that helps us navigate our lives or an AI-driven healthcare platform that anticipates our needs. Society will need to adapt, establishing new norms and frameworks for interacting with technology in ways that ensure **equity, inclusivity**, and **safety**.

The future of AI holds immense promise, and **Gemini 2.0** is at the forefront of this transformation. It is a glimpse into a world where AI can not only process information but also understand, reason, and act on that information in meaningful ways. Whether through **transforming industries**, **empowering education**, or **revolutionizing daily life**, Gemini 2.0 and other similar AI technologies will fundamentally change how we approach the challenges and opportunities of the future. As we move forward, it's essential that we continue to embrace these advancements with a thoughtful and responsible approach, ensuring that they contribute to a **better, more connected world** for all.

The journey of AI is just beginning, and its influence will continue to shape our world in ways we are only starting to understand. The future is undoubtedly exciting, and with Gemini 2.0 leading the way, we are poised to experience a world where human and AI collaboration takes us further than ever before. The question isn't whether AI will change the world—it's how we, as individuals and as a society, choose to navigate and shape that change. The power is in our hands.

Conclusion

As we bring this journey through the incredible world of **Gemini 2.0** to a close, it's clear that the developments we've explored are just the beginning of a much larger transformation in the realm of artificial intelligence. **Gemini 2.0** is a landmark achievement in AI evolution, offering groundbreaking capabilities that promise to reshape how we interact with technology and how we work across industries. Its **multimodal functionality**—allowing it to process and generate text, images, and even audio—opens up entirely new avenues for creativity, productivity, and innovation. From **enhanced speed and performance** to its **Agentic AI** capabilities, Gemini 2.0 is designed not just to respond but to anticipate and execute tasks with unparalleled efficiency.

Its integration with tools like **Project Astra** and **Project Mariner** takes the AI experience to new heights, offering real-world applications across industries such as healthcare, logistics, gaming, and even **AI-powered coding**. These advancements are setting the stage for a future where AI is not just a tool, but a partner—capable of working alongside humans to solve complex problems, enhance decision-making, and even create art. The potential impact of Gemini 2.0 is far-reaching, and it's only the beginning.

However, as we embrace these exciting advancements, we must remain vigilant about the ethical and societal challenges that come with them. As we've discussed, **privacy**, **security**, and **AI ethics** will play critical roles in shaping the future of AI, ensuring that the benefits are shared equitably and responsibly. Google's focus on **responsible AI development** with **Gemini 2.0** is a step in the right direction, but the path ahead will require ongoing dialogue, regulation, and careful consideration of how AI is integrated into society. We all have a stake in this revolution, and it's crucial that we stay informed and involved in the ethical debates and technological advances that will define the future of AI.

The future of AI is unfolding at an exhilarating pace, and the **Gemini 2.0** update is just one piece of the larger puzzle. The possibilities for AI to **transform industries**, **augment human potential**, and **reshape society** are virtually limitless. It's not just about developing better technology—it's about ensuring that this technology serves humanity, enhances our lives, and empowers individuals across the globe.

As we stand on the brink of this AI revolution, the most important thing we can do as individuals is to **stay curious**. Keep learning about these advancements, explore how AI can help solve problems in your own life, and engage in conversations about how to guide its development in a positive direction. The pace of innovation will only increase, and the more informed and involved we are, the better equipped we'll be to shape the future of this technology.

Stay engaged, stay curious, and most importantly—be ready. The world of AI is unfolding before us, and **Gemini 2.0** is just the beginning. The future is here, and it's up to us to shape it.

www.ingramcontent.com/pod-product-compliance
Lightning Source LLC
Chambersburg PA
CBHW071050240526
45469CB00006BD/2291